Acute Care –
Homeopathy for
Sciatica

By D C Rona, PhD, ND, DHM

Acute Care - Homeopathy for Sciatica

ISBN: 978-0-557-10560-1

Published by Lulu.com

Cover Photograph by Jessica Rona
Edited by Veneda B. Cook

Disclaimer

The information in this book is for educational purposes only and is not recommended as a means of diagnosing or treating any illness or condition. All matters concerning physical and mental health should be supervised by a health practitioner knowledgeable in treating that particular condition. Neither the publisher nor author directly or indirectly dispense medical advice, nor do they prescribe any remedies or assume any responsibility for those who choose to treat themselves.

Forward

Alternative, integrative, complementary – all words used to describe healing modalities outside the traditional 'modern' western medical framework. Many of these modalities have been used and proven effective for centuries, and some for thousands of years.

Most often, holistic healing methods are centered on individualization – where 'modern' medicine is based on diagnosis and protocols of treatment of a diagnosis and not an individual. Herein lies the great schism - the reason it is so often easy to dismiss the holistic modalities. It is far more difficult to statistically 'prove' the effectiveness of individualization than standardization. Statistics and standardization are perhaps a simpler approach to medicine, but not a substitute for holistic healing.

For 200 years, homeopathy has offered effective and safe results. It is fundamentally based on the body's inner abilities to heal and restore. True homeopathic healing results when the entire mind, body and spirit are considered in selecting remedies and potencies. However, it is also an exceptionally effective aide in acute situations – allowing the body to respond efficiently and allowing the whole system to work together to bring about a full recovery.

In the Acute Care series, the concepts and experience of applying homeopathy to a variety of acute situations is presented. Acute care is no substitute for the careful case taking and analysis of complete professional homeopathic care.

Homeopathy is completely complementary to traditional western modern medical care, and ideally should be integrated - using the best of each - for the care, healing and long term health of the individual.

Table of Contents

Chapter 1

What is homeopathy?

Throughout the history of mankind there have been attempts to understand the process of healing. The Law of Similars (like cures like) is found recorded in many places. A few examples include:

∞

"Through the like, disease is produced and through the application of the like is it cured."
Hippocrates (4ᵗʰ century BC)

∞

"You there bring together the same anatomy of the herbs and the same anatomy of the illness into one order. This simile gives you an understanding of the way in which you shall heal."
Paracelsus (5ᵗʰ Century AD)

∞

Samuel Hahnemann (1755-1843) was the founder of the system we call Homeopathy. Rediscovering the Law of similar, which he phrased as *Similia Similibus Curentur,* which translates as *"let likes cure likes"*, he developed a system using infinitesimally small doses to stimulate the organism to heal itself. He termed this medical system *Homeopathy* from the Greek words *homoios* for "similar" and *pathos* for "suffering" or "disease". Hahnemann's system is based on the principle which states that any substance, which can cause specific symptoms when given to a group of healthy people, can help to heal those who are ill and experiencing similar symptoms. The proper medicine, in very small amounts, promotes the reaction of healing within the body. In fact, he found that the smaller the dose, the more positive the cure.

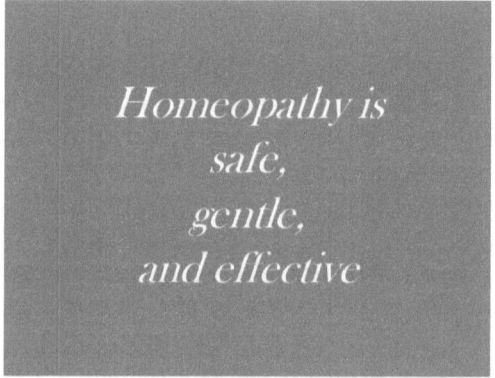

Homeopathy is safe, gentle, and effective

In 1810 Hahnemann wrote "Organon of Medicine". In subsequent revisions, he continued to refine and develop the system of homeopathy. Homeopathy became 'mainstream' medicine in the United States in the early 1800's. Numerous medical schools and hospitals still bear the name of Hahnemann. The first national medical association in the United States was the American Institute of Homeopathy, (founded in 1844). Politics (and not efficacy) prevailed in favor of pharmaceuticals and surgeries as a more 'modern' approach to medicine through most of the 20th century.

Because homeopathy works to support the vital force of the body, it assists the natural ability of the body to heal faster and does not suppress symptoms or delay the healing activities of the body. This is in contrast to pharmaceuticals that work by suppressing symptoms, even the symptoms that are healing responses (for example, acetaminophen suppressing a fever). Pharmaceuticals acting in this way can actually cause the body to take longer to recover from an acute injury (such as a sprain) or an infectious illness (such as cold or flu). Also, homeopathic remedies do not produce side effects (i.e. drowsiness) and they may be safely taken in conjunction with conventional medicines or treatments when necessary.

> *Homeopathy is the second most popular therapy in the world with over 300 million patients worldwide in 65 countries according to the World Health Organization (WHO).*

With an unrivaled safety record over the past 200 years, and a record of true effectiveness, it is easy to see why homeopathy is considered a mainstream medical therapy in Europe.

The concept of an individual's "constitution" is common to many systems of healthcare, including Homeopathy, Traditional Chinese, Indian Ayurvedic, and European Naturopathic. It holds that the individual has certain trends that are part of the basic nature of that individual. While one may thrive in open air and crave fatty foods, another feels best in cold and craves solitude and sweets, while another feels oppressed by open environments and seeks warmth.

Of course, these examples only touch at the very tip of the entirety, but they serve to illustrate the concept. Each of these systems of medicine has a procedure for denoting and clustering physical, mental, emotional, and spiritual traits into constitutional patterns. By perceiving these patterns, one may be more effective in finding remedies that best fit that individual.

In homeopathy, the individual's constitution is identified after a lengthy interview process (and sometimes amended over time as the information is provided). This information is used to refine the search for the specific remedy that best fits the entire picture of the individual. Most often this information is used in treating long term disorders, but it may be incorporated into the choice of remedy for acute illnesses and even treatment after an accident. This type of investigation, analysis and remedy selection is part of 'classical' approach, adhering strictly to the original principles laid down by Hahnemann. By contrast, there are many versions of prescribing homeopathic remedies varying from using electrical devices to select remedies, to patterns of protocol for remedies by disease, even injecting solutions containing homeopathic materials – all of which are non-classical and non-Hahnemannian.

In working with a homeopath it's important to pay attention to all the reactions of the body. In relating information on any symptom, such as an aching back or sore throat, as, it is important to note specific details, such as:
- When did the symptom begin – and what were the circumstances
- An accurate description of the type of sensation (i.e. stabbing pain as opposed to burning pain)
- The exact localization (i.e. did the pain begin on the right side and move to the left, bore inward, etc.)
- The extension or movement of the pain (does it radiate up the foot, down the back, etc.)
- Does the symptom reoccur at a set interval of time?

- What makes the symptom better or worse (i.e. cold, heat, water, etc.)
- Is the symptom accompanied by any psychological change (i.e. sad, giddy, melancholy, etc.)
- Since the onset, has there been a craving for food or water (i.e. salt, sugar, cold water, etc.)

These are the details that will allow the homeopath to identify the correct remedy for the individual.

> *Homeopathy provides safe, natural effective remedies and individualized therapy, which encourages a cooperative relationship between patient and doctor.*

Complementing mainstream 'traditional' research studies, two centuries of practice have shown the effectiveness of homeopathic medicines in treating a variety of musculoskeletal conditions including trauma, joint and muscle pain, fractures and sprains, and sciatica and lumbar pains. Depending on your activities and lifestyle, you may (over a lifetime) experience sport related injuries, home injuries, automobile accidents, or even industrial accidents. The basic application of homeopathic principles and remedies will be the same. Choice of homeopathic medicine as well as the frequency of dosage and the potency vary according to the seriousness of the injury and its acuteness. Old injuries which have become chronic will be treated differently than recent ones. The beauty of these medicines is in the value of the deep tissue, bone and nerve healing as well as pain relief. The homeopathic medicines used for acute injuries should be used in conjunction with conventional diagnostic tests (x-rays) treatment methods (splinting, casting, etc.) as well as basic care such as rest, ice packs, compression and elevation of the affected area.

> *When using homeopathy directly on the soccer field, I find that the results are almost immediate"*
>
> Dr. Jean-Marcel Ferret
> Doctor to the French soccer team
> from 1993 to 2004

For those individuals who might be prone to frequent acute sciatica episodes – you may wish to work with a homeopath to select homeopathic remedies for immediate 'first aid' support until you can reach appropriate medical care. A skilled homeopath can help you select remedies that best fit your individual symptoms.

Homeopathic practices

Homeopathy is practiced as a complementary and/or alternative form of medicine. Homeopathy is frequently incorporated into the daily practice of other healing modalities including chiropractic, naturopathic, and sports physical therapist. In Europe, it is common to see homeopathy included in traditional western medical practices as an important modality. Acute application of homeopathic remedies requires minimum training and experience to achieve excellent results. However since several homeopathic medicines can address the same type of condition, anything beyond acute/emergency level treatment must take into account all of the particular symptoms of the individual. Appropriate education and training is necessary to utilize homeopathic remedies to their full potential. To work with any chronic condition requires full individualization, properly taking the case (history as well as details of symptoms and sensations) and analyzing not only the proper remedy but also the proper dosing. There are many schools of homeopathy throughout the world. For a listing in the United States and Canada, visit the website for the National Center for Homeopathy:

(http://nationalcenterforhomeopathy.org)

for schools, practitioners and other resources.

The jargon of homeopathy

Aggravation – the worsening of a symptom. This may be temporary homeopathic aggravation as the individual is healing, or disease aggravation if the underlying disease state worsens.

Constitution – the sum of the patient's symptoms, appearance, temperament, etc. that make it possible to select a remedy that fits their totality

Homeopathy – from the Greek homoios (like) and pathos (suffering) Homeopathy is a branch of medical therapeutics where the prescription of remedies is derived from the pharmacological principle of analogy (like cures like) and applied with small or infinitesimal doses. Samuel Hahnemann described this principle by using a Latin phrase Similia Similibus Curentur, which translates as "let likes cure likes." While known for centuries, Hahnemann applied this principle in developing a system of medicine called homeopathy. It has been used successfully for the last 200 years.

Materia Medica – a listing of the remedies with their corresponding symptom pictures both from proving and clinical experience

Modality – the factors which influence a symptom. These are the factors that make the symptoms worse or better for that individual, or the way in which a symptom moves

Polychrest – term given to 30-45 remedies that are used very frequently because they have such a wide range of applications

Potency – the strength of a remedy determined by dilution and succussion and denoted by X, C, M or LM depending on the dilution ratio used in preparation. The higher the dilution the 'stronger' or more potent the remedy.

Proving – experiments carried out by healthy volunteers to determine the symptom picture produced by each remedy

Remedy – homeopathic remedy – homeopathic medicine are terms used interchangeably in the discussion of homeopathic treatment and are used to describe a diluted and succussed medicinal substance which is 'proven' and included in the Homeopathic Pharmacopia.

Repertory – a cross reference system of indexing the symptoms (rubrics) produced by remedies

Vital force – the energy responsible for the state of health of an individual

Chapter 2

What constitutes an acute?

Dictionary definitions are simple black and white definitions. The reality of human experience is not so simple. Certainly there is a wide spectrum of situations. The car accident is the most simplistic illustrations of a pure acute situation – where the previous state of health or underlying past injuries did not contribute to the situation. A birth defect limiting kidney function in an adult is a simplistic illustration of a pure chronic situation. Most acute symptoms fall somewhere in between. The intense pain of a weekend athlete who's years of past injuries and

Acute
(from the Latin acutus, past participle of acuere – to sharpen, from acus – needle)

- Characterized by sharpness or severity (acute pain)
- Having sudden onset, sharp rise, and short course (acute disease)
- Being, providing, or requiring short term medical care (as for serious illness or traumatic injury)
- Lasting a short time

Chronic
(from the French chronique, from Greek chronikos of time, from chromos)

- Marked by long duration or frequent recurrence
- Always present or encountered, constantly vexing, weakening or troubling
- Being habitual

First Aid

Emergency care or treatment given to an ill or injured person before regular medical aid can be obtained

poor training regime produce 3 or 4 episodes of back pain per year could be considered an acute phase of a chronic problem. Likewise, severe muscle cramps could be an acute phase of a chronic disease.

Within homeopathy, it is common to hear a general classification of acute vs. chronic diseases in discussing illnesses. Acute disease is usually defined as self-limited and brief, generally running their course within a reasonable period of time (i.e. colds, flu or viruses). Chronic disease is the term used for the situation where in individual suffers from repeated attacks of the same illness or the body presents a building set of symptoms. Acute diseases are often self treated or require only limited training to handle very well with homeopathy. Any time the situation is judged to be chronic; it requires study and diagnosis by a homeopathic physician to be fully resolved (i.e. asthma, migraines and stomach disorders) without concern for suppression of symptoms.

Why is this important? Certainly homeopathic remedies may be applied to any acute situation, even those that are acute phases of chronic problems. However, if a chronic problem is the root of

a current acute symptom – the individual will be best treated by fully individualized homeopathic treatment.

For the purpose of this book, we will focus on the use of homeopathic remedies for the typical acute situations that are seen in the course of normal day to day life - the intense pain and limited mobility that accompany sciatica. Also, for the purpose of this book, we will focus on first aid assistance, assuming that appropriate, prompt and best possible medical diagnosis and treatment will be pursued or is taking place concurrently with the use of homeopathic remedies.

Homeopathic remedies are fast to act when used in acute situations and are helpful in pain relief and promoting the body's best healing responses. When homeopathic remedies are used in conjunction with conventional first aid procedures, the risk of long-term damage from an injury can be significantly reduced and the healing process is noticeably enhanced.

Homeopathy's greatest advantage as a component of acute "first aid" care is that it has no negative side-effects. It will not interfere with necessary traditional medical treatments and will enhance the body's ability to withstand the stress of procedures (such as the setting of a bone) and will usually significantly shorten the recovery time.

It is not uncommon to hear treatment and recovery defined in terms of phases or stages. Within different medical disciplines this breakdown will vary. In accidents, it is often broken into Phase 1 – the first 24 hours, and Phase 2 – the time after phase 1 in which the original symptoms persist. These descriptions are arbitrary, and while they may be communication tools, they should not define a treatment protocol – since every individual must be evaluated as an individual and not some standardized protocol.

In general, in acute situations, well chosen homeopathic remedies should bring a sense of relief within a matter of minutes, with substantial relief felt within a matter of hours. The ability to sense relief depends on the individual, their current state of overwhelm in pain, their perception of fear over the situation, etc. The remedy may need to be repeated several times over the short term to achieve and hold the sense of relief. Of course, true healing takes time. The benefit of homeopathic remedies is that they not only have provided comfort and relief in the acute situation, but also they are supporting rather than interfering with the body's natural healing process, so recovery may begin immediately. Returning to health and vigor is of course the primary objective.

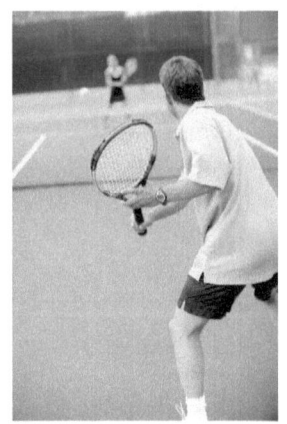

Chapter 3

Homeopathic remedies

Unlike some of the products found in health food stores - homeopathic medicines are made according to strict pharmaceutical and regulatory requirements. There are relatively few manufacturers of homeopathic remedies. A multidisciplinary team of pharmacists, chemists, zoologists and botanists must monitor every stage of homeopathic medicine manufacture.

The raw materials may come from animal, vegetable or mineral sources. They are screened and used in compliance with current pharmaceutical regulations, homeopathic tradition and official Pharmacopeia.

While the Europeans and Americans have slightly different 'standards books', any company preparing homeopathic remedies for sale in Europe or the United States must meet stringent requirements. The Homeopathic Pharmacopoeia Convention of the United States is responsible for establishing good manufacturing procedures and industry standards in collaboration with the FDA. In the US, the Food and Drug Administration (FDA) has recognized the Homeopathic Pharmacopeia of the United States (HPUS) since 1983. As an official reference on the subject, it enables homeopathic stocks to be registered. In order

to be placed on the American market, the homeopathic medicine must meet the conditions established by the FDA in 1998.

These homeopathic medicines are most often found on the store shelves packaged in granule tubes or globule doses. The vocabulary of the pill sizes varies widely depending on the training of the individual or the country of origin of the medicines. In homeopathic references you will see the tiny pills described as pills, pillules, granules, globes and tablets. There are also other forms such as pomades, gels, syrups, caplets, eye washes, drops, suppositories, etc.

Remedy names are given in Latin, which provide the most consistent means for accurately identifying the remedy. You may note that homeopathic names may differ from common names that are used when in herbal form. A good example of this is the homeopathic remedy Crataegus which is more commonly known as Hawthorn.

There will be abbreviations used on labels, simply because they run out of room. An example is Rhus Toxicodendron (poison ivy) which you will often see abbreviated as Rhus tox.

A common error in referring to homeopathic remedies comes when only part of a name is used. For example, you will frequently see reference to Aconite or Aconitum. However, Aconite is available in a variety of forms including:

> Aconitum cammarum
> Aconitum ferox
> Aconitum lycotonum
> Aconitum napellus

Normally it is Aconitum napellus that is available in local health food stores. The other varieties would most likely need to be special ordered. However, when looking at a remedy name in a reference book, such as a Material Medica or on a label – it is important to ensure you are looking at the full name of the remedy.

The potency number is found on every remedy, immediately following or below the remedy name. Homeopathic medicine consists of natural substances that are extremely diluted. The amount of dilution is called potency. In addition to dilution, remedies are also 'succussed' in the process. Succussion is shaking with impact (the early preparations were made by holding the dilution bottle in the hand and slamming the bottle/hand onto a book), which has been shown to enhance the therapeutic effect of the remedy.

The label will contain the remedy name(such as Arnica montana or Apis mell.) and a potency number (such as 30C or 30X). There will also be one or two self limiting conditions for which this remedy may be used (required for an over the counter item).

The "X" after the potency number (as in 6x) refers to the number of times in which a medicine is diluted 1:10, while the "C" after the potency number (as in 6c) is diluted 1:100. There is also an M scale which is an extension of the C scale. The higher the potency used, the more accurate the remedy selection must be. Also the higher the potency, the more often the psychological factors are involved and affected. Because of this, it is usually recommended to use only up to the 30th potency unless you are an experienced and well trained homeopath. Additionally, 200 years of homeopathic clinical experience has found that between the 6th and 30th potency are exceptionally effective for acute physicals, and they may be repeated more often. Some homeopaths contend that the lower potencies have a greater affinity for physical trauma. Emergency treatment most often utilizes 6x, 6c, 12x, 12c or 30c.

In his later years, Dr. Hahnemann developed the LM scale of potencies, using dilutions on the 1:1000 scale. These were by far his most potent and yet gentlest remedies. These are described in his final (6th) edition of the Organon of Medicine. Many of the most effective homeopaths in the intervening 200 years did not know of this last development since the 6th edition was not published immediately. Now, due to the exceptionally effective legacy of those 4th and 5th edition practitioners, and due to the nature of the dosing of the LM potencies, most remedies available in our neighborhood health food stores are X or C potencies.

> *Note: In Europe the D (decimal) scale is often utilized. A wide range of remedies is on hand at almost every neighborhood pharmacy, and often prescribers will use a greater variety of potencies than we find readily available in the US.*

Low potencies (i.e. 3x, 6x, 6c, 12c) are generally used for more localized issues and for acute situations. Higher potencies (30c and higher) are generally used when the patient's general health is a factor or where psychological influences are important to the case. These simple rules of thumb are just that – and many times an experienced practitioner will use potency based on additional factors such as the general health and energy level of the patient. Higher potencies are best prescribed by the more experienced homeopathic practitioners since the accuracy of the remedy choice becomes more critical. Low potencies are often repeated more frequently, as the patient feels it necessary, while higher frequencies are not repeated often if at all. In all use of homeopathy, the accurate selection of the remedy is far more important than the accurate selection of potency.

Single Remedies and Combination Formulas

Homeopathic medicines are available as single remedies or as formulas of two or more remedies mixed together. Herein lies great contention between those classically trained homeopaths and those who have come to use homeopathy as a 'this for that' treatment - which is an allopathic approach that simply uses the excellent effects of homeopathic remedies. The classically trained homeopath will stand firm that the ONLY way to treat homeopathically, using homeopathic remedies, is to individualize the single remedy that most clearly fits the presenting complaints of the individual. Anything other than complete individualization is not homeopathic – since at its very core, classical homeopathy is TOTALLY individualized. Others who have found homeopathic remedies to be exceptionally useful, but do not have the time or training to completely individualize, have found that it can be effective to mix a few remedies that are 'usually' applicable to a particular situation, and give that mixture to anyone in a specific situation. You will find even highly reputed homeopathic manufacturers that will put out a line of these combination remedies in addition to the single remedies since they are effective, and they meet the needs of a segment of the population.

Then there are a few individuals who have pushed that boundary even further and put a mixture of tens of remedies into a combination, and still contend that they are using 'homeopathy'.

Single remedies are more effective for any situation when you determine the correct medicine. By giving a single remedy, you have the option to use the precise potency (or change the potency) to meet the needs of the individual, or change the remedy as the symptoms change. Combination or 'formula' products usually contain remedies in the 3, 6, or 12th potencies. Severe pain in otherwise healthy and vigorous individuals may respond more rapidly by selecting from 30 C or 200 C potencies. When combination remedies work, you do not know which component was working, so if you desire to continue treatment and need to change potencies, you do not know which single remedy to use. When combination remedies do not work, you have wasted precious healing time, and have no idea if it were the individual remedy components that were not effective, or if one worked well and the others neutralized its effect.

Remedy Information Resources

Homeopathic remedy information is indexed into two types of references; Materia Medica and Repertories. The Materia Medica is an organized description of each homeopathic remedy. Depending on the individual Materia Medica, you may find a quick short 'snapshot' of the remedy or you may find encyclopedic coverage. To be useful to a professional homeopath, the Materia Medica will give various characteristics and symptoms a person might exhibit who needs that specific remedy. Also commonly listed are clinical uses for the remedy, keynote symptoms, modalities and proving symptoms.

Modalities are listings of what makes someone feel better or feel worse. For instance, Bryonia will list the modality of worse with motion, whereas Rhus tox will list the modality of worse with inactivity. Correctly applying these modalities in the search for the best remedy for an individual is straightforward, but a learned and practiced art.

Homeopathic remedies go through a process of "proving" to determine the curative properties of each remedy. In a proving, health humans are given the item and the symptoms produced are carefully recorded. Homeopathy's most basic premise is Like Treating Like or Similar Treating Similar. So symptoms produced in a proving are the same symptoms that can be treated with that remedy.

The following example of a very short Materia Medica for Arnica is taken from *Boericke's Materia Medica,* 1901, written by William Boericke.

Arnica montana (*Leopard's Bane*)

Produces conditions upon the system quite similar to those resulting from injuries, falls, blows, contusions. Tinnitus aurium. *Putrid phenomena.* Septic conditions; prophylactic of pus infection. Apoplexy, red, full face. It is especially suited to cases when any injury, however remote, seems to have caused the present trouble. *After traumatic injuries,* overuse of any organ, strains. Arnica is disposed to cerebral congestion. Acts best in plethoric, feebly in debilitated with impoverished blood, cardiac dropsy with dyspnoea. A muscular tonic. Traumatism of grief, remorse or sudden realization of financial loss. Limbs and body ache as if beaten; joints as if sprained. Bed feels too hard. Marked effect on the blood. Affects the venous system inducing stasis. Echymosis and haemorrhages. Relaxed blood vessels, black and blue spots. Tendency to haemorrhage and low-fever states. Tendency to tissue degeneration, septic conditions, abscesses that do not mature. *Sore, lame, bruised feeling.* Neuralgias originating in disturbances of pneumo-gastric. Rheumatism of muscular and tendinous tissue, especially of back and shoulders. Aversion to tobacco. *Influenza.* Thrombosis. Hematocele.

Mind.--Fears touch, or the approach of anyone. Unconscious; when spoken to answers correctly, but relapses. Indifference; inability to perform continuous active work; morose, delirious. Nervous; cannot bear pain; whole body oversensitive. Says there is nothing the matter with him. Wants to be let alone. Agoraphobia (fear of space). After mental strain or shock.

Head.--*Hot, with cold body*; confused; sensitiveness of brain, with sharp, pinching pains. Scalp feels contracted. Cold spot on forehead. Chronic vertigo; objects whirl about especially when walking.

Eyes.--Diplopia from traumatism, muscular paralysis, retinal haemorrhage. Bruised, sore feeling in eyes after close work. Must keep eyes open. Dizzy on closing them. Feel tired and weary after sight-seeing, moving pictures, etc.

Ears.--Noises in ear caused by rush of blood to the head. Shooting in and around ears. Blood from ears. Dullness of hearing after concussion. Pain in cartilages of ears as

if   bruised.
Nose.··Bleeding after every fit of coughing, dark fluid
blood.   Nose   feels   sore;   *cold*.
Mouth.··*Fetid breath*. Dry and thirsty. Bitter taste
(*Colocy*). *Taste as from bad eggs*. Soreness of gums after
teeth extraction (*Sepia*). Empyaema of maxillary sinus.
Face.··*Sunken*; very red. Heat in lips. Herpes in face.
Stomach.··Longing for vinegar. Distaste for milk and
meat. Canine hunger. Vomiting of blood. Pain in
stomach during eating. Repletion with loathing.
Oppressive gases pass upward and downward. Pressure
as from a stone. *Feeling as if stomach were passing
against   spine.   Fetid   vomiting*.
Abdomen.··Stitches under false ribs. Distended;
offensive flatus. Sharp thrusts through abdomen.
Stool.··*Straining of tenesmus in diarrhoea. Offensive*,
brown, *bloody*, putrid, involuntary. Looks like brown
yeast. Must lie down after every stool. Diarrhoea of
consumption; worse lying on left side. Dysenteric stools
with   muscular   pains.
Urine.··Retained from over-exertion. Dark brick-red
sediment. Vesical tenesmus with very painful
micturition.
Female.··Bruised parts after labor. Violent after-pains.
Uterine haemorrhage from mechanical injury after
coition. Sore nipples. Mastitis from injury. Feeling as if
foetus   were   lying   crosswise.
Respiratory.··Coughs depending on cardiac lesion,
paroxysmal, at night, during sleep, worse exercise.
Acute tonsillitis, swelling of soft palate and uvula.
Pneumonia; approaching paralysis. Hoarseness from
overuse of voice. Raw, sore feeling in morning. Cough
produced by weeping and lamenting. Dry, from tickling
low down in trachea. Bloody expectoration. Dyspnoea
with haemoptysis. All bones and cartilages of chest
painful. *Violent spasmodic cough, with facial herpes*.
Whooping cough, child cries before coughing.
Pleurodynia   (Ranunc;   Cimicif).
Heart.··*Angina pectoris*; pain especially severe in elbow
of left arm. Stitches in heart. Pulse feeble and irregular.
Cardiac dropsy with distressing dyspnoea. Extremities
distended, feel bruised and sore. Fatty heart and
hypertrophy.
Extremities.··Gout. Great fear of being touched or
approached. Pain in back and limbs, as if bruised or
beaten. Sprained and dislocated feeling. Soreness after

overexertion. Everything on which he lies seems too hard. Deathly coldness of forearm. Cannot walk erect, on account of bruised pain in pelvic region. Rheumatism begins low down and works up (*Ledum*). **Skin.**··*Black and blue*. Itching, burning, eruption of small pimples. *Crops of small boils* (*Ichthyol; Silica*). Ecchymosis. Bed sores (*Bovinine locally*). Acne indurata, characterized by *symmetry in distribution*. **Sleep.**··Sleepless and restless when over tired. Comatose drowsiness; awakens with hot head; dreams of death, mutilated bodies, anxious and terrible. Horrors in the night. Involuntary stools during sleep. **Fever.**··Febrile symptoms closely related to typhoid. Shivering over whole body. Heat and redness of head, with coolness of rest of body. Internal heat; feet and hands cold. Nightly sour sweats. **Modalities.**··*Worse*, least touch; motion; rest; wine; damp cold. *Better*, lying down, or with head low. **Relationship.**··Antidotes: *Camph. Vitex trifolia.* ·Indian Arnica (Sprains and pains, headache in temples, pain in joints; pain in abdomen; pain in testicles). Complementary: *Acon; Ipec.* Compare: *Acon; Bapt; Bellis; Hamam; Rhus; Hyperic.*

Homeopathic vocabulary can sometimes appear confusing. The original works of Hahnemann and the early homeopaths were written in German, many of the early translations used a combination of terms common to the practice of medicine in that time period. Translations will often contain archaic sounding descriptions, but they are no less valuable. For example, feeling nauseous when riding in a carriage is completely equivalent to motion sickness from today's car, plane or train.

Repertories are a cross reference to remedies where the symptoms or illnesses are grouped by body part, and the remedies known to have those particular symptoms are listed.

The following short example of a repertory listing is from
THE REPERTORY by Oscar E. Boericke, M.D.

From · LOCOMOTOR SYSTEM

SHOULDERS-SCAPUAE

Deltoid, pain, rheumatism ·· Ferr. p., Glycerin,
Lycopers., Med., Nux m., Ox. ac., Rhus t., Sang., Sticta,
Syph., Urt., Viola od., Zinc. ox., Zing.

PAINS ·· Alum., Am. phos., Anag., Arn., Azadir., Bar.
c., Can. ind., Chel., Cocc., Con., Fagop., Jugl. r., Kreos.,
Lyc., Menisp., Myr., Nat. ars., Nat. c., Nit. ac., Ran. b.,
Sep., Ver. a., Viscum.

Scapulæ [between] ·· Æsc., Ars., Bry., Camph., Chenop.,
Euonym., Granat., Guaiac., Mag. s.

Burning, in small spots ·· Agar., Phos., Ran b., Sul.

Drawing ·· Ars., Berb. v., Cham., Col., Sul.
Left [in] ·· Acon., Æsc., Agar., Ant. t., Aspar., Cham.,
Chenop. gl., Col., Eup. purp., Ferr. m., Led., Lob. syph.,
Nux m., Onosm., Rhodium, Stram., Sul.

Lower angle ·· Chenop. gl., Cupr. ars., Ran. b.

Right [in] ·· Abies c., Am. m., Bry., Chel., Chenop. gl.,
Col., Ferr. p., Ferr. mur., Guaco, Ichthy., Ipomuea, Jugl.
c., Kali c., Kal., Mag. c., Pall., Phyt., Puls. nut., Ran. b.,
Sang., Solan. lyc., Sticta, Stront., Urt.

Lower angle ·· Chel., Chenop. anth., Kali c., Merc., Pod.

Rheumatic ·· Acon., Am. caust., Berb. v., Bry., Colch.,
Ferr. mur., Ferr. p., Guaiac., Ham., Kali c., Kal., Lact.
ac., Led., Lith. c., Lith. lact., Med., Ol. an., Pall., Phyt.,
Prim., Radium, Ran. b., Rhod., Rhus t., Sang., Stellar.,
Sticta, Stront., Sul., Syph., Urt., Viola od.

Stitches, tearing ·· Am. m., Bry., Hyper., Kali c., Lyc., Mag. c., Nit. ac.

Stiffness ·· Cocc., Dulc., Granat., Indium, Phyt., Prim. v., Sang., Senec. jac

Expiration date for homeopathic remedies

You will usually see an expiration date on homeopathic labels of approximately 5 years from date of manufacture. There is no reason (other than labeling laws) for this date since remedies made 200 years ago are still useful. The usual warnings to keep remedies in cool, dry places out of direct sunlight are wise for any medicine. Most also consider it wise to keep homeopathic remedies away from intense magnetic fields and intense substances such as mothballs, essential oils, etc.

Sources for homeopathic remedies

Many health food stores carry some homeopathic remedies – and most often they will carry all of the basic 'first aid' remedies in 12c or 30c. This is adequate for most situations. Most tubes will contain roughly 80 pillules (depending on the size of the pillules), easily fits in the pocket or purse, and will usually cost between $5 and $9 depending on the store. If you are constructing an emergency first aid kit for the office, a sport team field kit, hiking or travelers kit, or a home disaster kit you may wish to use a homeopathic pharmacy (either in person or online) to obtain the remedy selection and potencies that best match your needs. A professional homeopath can help you design a kit to match your specific needs. Then use a search engine for "Homeopathic Remedy Online Store" to get a current listing of online sources.

Chapter 4

When, which and how to use homeopathic remedies

Choosing the Remedy

Selecting the exact remedy to meet the full needs of an individual requires significant training and experience and an in depth interview to ascertain the full picture. Selecting an excellent first aid remedy to meet the needs of an acute situation is relatively easy and will significantly aid in immediate comfort and long term healing of the acute. Each remedy has a profile of the physical and psychological aspects that it covers. These profiles are found in a Materia Medica and can be voluminous, truly encyclopedic detailed portraits. Each remedy also has a shortened set of 'key notes' which can give you a quick 'snap shot' of its uses. These key notes are something like an outline or short synopsis of a massive book. You might use such an outline to decide between War and Peace and The Iliad and Odyssey, but it would not provide anything of the true depth, color and quality of the books. Key notes are wonderful to use in emergencies, but any time you can, using the full Materia Medica will give a better understanding of the remedy potential. Keynotes may occasionally be misleading. For example, a keynote for a remedy may be 'lack of thirst' but the remedy may be exceptionally useful for an individual feeling great thirst on some occasions. Overall, you are very safe in using keynotes for first aid situations.

For acute treatment at home, or out hiking, biking, or coaching soccer, a small number of well documented acute remedies may be used with excellent results. In

this case, you only need a short description of each remedy to make a good selection. You will find this sort of information in Section 5. Should you find these remedies do not appropriately cover your needs or the events you encounter most often, you may create your own quick reference by consulting with a professional homeopath.

Taking a Remedy

It is very convenient to keep the small tubes or bottles of remedies in a first aid kit. Usually to dispense from the small tubes you hold the tube upside down and twist the plastic top until 2 or 3 pillules are in the clear plastic top. You then pop the pillules directly under the tongue and let them dissolve. With the small bottles, you tip 2 or 3 pillules into the top or onto a clean spoon and again, pop them under your tongue to dissolve. This is a very effective dosing technique and is called 'dry dosing'. Larger bottles or jars of remedies require the pills or tablets be dispensed into a spoon, paper, or other such method prior to placing them in the individuals mouth – if they are to be taken dry. Professionals may dispense a single dose (2 to 4 pillules) placed in a neatly folded paper or small envelope with instructions to place the pillules under the tongue at a specified time (without touching the pillules).

When possible, the remedy should be placed in water (spring or distilled is preferred over tap water) and sipped. This is commonly referred to as 'wet dosing'. There is an entire methodology of dilution and 'wet' doses within homeopathic practice, but for acute care, simply dissolving the pills into water will greatly speed the action of the remedy. It also permits you to alter the potency between dosing by shaking or stirring the remedy solution. This minor change permits the best possible healing potential.

For ongoing treatment of an acute, it is common to prepare a small dropper bottle (30 ml or smaller) so the individual may continue to take doses through the healing process. To prepare such a bottle the instructions may call for 3 to 10 pillules into distilled or spring water. Drops are then dispensed directly into the mouth or into a glass or water at a prescribed timing, or as needed, with several intense shakes between doses. This dilution may also be applied to the temples or wrist in some situations.

If the bottle is kept for more than a day or so it is appropriate to add a drop or two of vinegar or brandy to aid in preservation.

During competitions and challenging training sessions athletes sometimes add Arnica to the regular water bottle and sip throughout the competition or training session. In these situations, low potencies (under 30c) are usually selected. After a major trauma a professional homeopath might suggest higher potencies such as 200c or 1M be used for severe injuries and pain. This requires accurate remedy selection.

Avoid Contamination
Homeopathic remedies are remarkably stable and forgiving, but you can take steps to keep them at their most effective. Rough handling with dirty fingers can potentially contaminate the remedies and it is always advised not to directly touch or handle the dry remedy pillules – but rather take them from the dispensing cap, a clean spoon, paper, etc. If you spill the remedies, do not put them back in the bottle.

To keep remedies effective for years - avoid heat above 120 degrees, moisture, and extended periods of sunlight. It is best to store at room temperature in a drawer or box, away from chemicals and electromagnetic fields. Dry remedies should not be stored in the refrigerator; however wet remedies may be stored in the refrigerator.

Dosing strategies
We are used to standardized dosing protocols in the traditional pharmacological industry. One size fits all. In homeopathy, dosing instructions vary widely. An experienced prescriber will evaluate the situation, the sum of the individual's injuries, the overall 'vital energy' of the individual, and the events yet to be faced by the individual (i.e. medical procedures, wait times in the emergency room, etc.) before determining a dosing

strategy, and even then, it will change with feedback from the injured individual. In most emergency situations, the dosing begins at a low potency (usually 6x, 6c, 12x, 12c, 30x or 30c) every 15 to 20 minutes for severe pain. As slight improvement is noticed, the frequency is reduced, usually to every 1 to 4 hours, or stopped altogether. The remedy is repeated then only as needed. Rarely do homeopathic remedies need to be repeated after the first 24 hours. [note: chronic conditions take longer.] The exception to this is a combination of physical re-injury (i.e. setting of a broken bone), chemical antidoting (i.e. pharmaceuticals), emotional aggravations (if you have been in a hospital recently you know what these are), and other forms of antidoting of the remedy (see the next section). Very rarely an individual may experience a homeopathic aggravation (an overstimulation of the healing process) after taking a remedy – in which case once the remedy is stopped, the aggravation will subside, and you may consider retaking the remedy – usually at a lower potency. Homeopathic aggravation is exceptionally rare in emergency situations with appropriate acute remedies.

Very rarely two remedies are well indicated for the situation. If that is the case, best results are found by waiting about 15 minutes between taking each remedy. Switch remedies if the symptoms change and begin to show that another remedy better fits the situation. Discontinue all homeopathic remedies when improvement is stable and steady. Let the body do its job, and support with nutrition, water, rest, etc.

The most important factor of homeopathic treatment is the selection of the correct remedy. No matter what potency the remedy, if it is the right remedy, it will have a healing effect on the body. If taken properly, no side effects or harm should ever occur. Remember, it is seldom necessary to take a homeopathic remedy for a long period of time.

Antidoting

It is possible to neutralize the effect of a homeopathic remedy and render it useless. Remedies should be kept away from airborne contaminants such as perfumes, aromatic oils and camphor. Smoking and caffeine can affect the effectiveness of remedies. Mint is generally considered to neutralize or counteract the action of homeopathic remedies. It is wise to not use any stimulants for two hours before and after taking homeopathic remedies, if possible. The following could possibly antidote the remedy:

- coffee
- mints
- strong smelling linaments, e.g. tiger balm, vicks vapo rub, eucalyptus, or camphor,
- perfumes & colognes

In an emergency situation – just give the remedy and don't worry about anything else. If you find the remedy is not acting effectively, and you are confident it was the right choice, then look at the possible antidoting factors and re-dose.

External Applications

While we most often think of homeopathic remedies as 'internal' remedies, there are several homeopathic remedies that are available for topical applications. Most often you will find Arnica, Calendula or Hypericum in creams, ointments, gels and even sprays. They are effective and easily used. Select the remedy based on the same information you would use to select internal remedies – and select the vehicle based on the specific situation. Ointments are usually placed in a petroleum base, which means they are not easily washed or wiped off, but do not allow the skin to breath. Gels and sprays allow the skin to breath but do not remain long on the skin of an active individual.   Creams are usually comforting and soothing so are more often used at bedtime.

Several additional factors must be considered with external applications. Calendula is excellent for healing cuts and scrapes, and arnica should not be used with any open skin. These specific directions will be found on the packaging for each product.

Suggestions for Taking Homeopathic Remedies
(under ideal situations)

- Take nothing by mouth 10-20 minutes prior to or following dosage. This includes food, drink, chewing gum, mouthwash, toothpaste etc.

- Limit as much as possible caffeine in any forms, such as soft drinks, coffee, or chocolate.

- No camphor, such as in muscle and joint rubs (e.g. sports balm, vapo rub, eucalyptus creams)

- Avoid moth ball fumes.

- Avoid perfumes and essential oils (this includes strong smelling household cleansers)

- Avoid dental drilling.

- Limit contact with cigarette smoke, especially menthol as much as possible.

- Avoid raw garlic, cooked is okay.

Chapter 5

Selected remedies for acute care of sciatica

While there are literally hundreds of homeopathic remedies that could be appropriate for consideration in the care of an individual with symptoms of sciatica (or sciatic neuritis), we will look at a selection of 19 remedies that are exceptionally useful for acute care. Remember, homeopathic remedy selection is based solely on the symptoms of the individual and not the diagnosis.

Acute sciatic pain may be from one or more root causes. In fact, the term sciatica or sciatic neuritis is a designation of symptoms, not a specific diagnosis for what is causing the symptoms. These basic structural causes of

Sciatic nerve

compression or irritation include:

- compression of lumbar nerves L4 or
  L5 or sacral nerves S1, S2 or S3
- compression of the sciatic nerve itself
- spinal disc herniation
- roughening, enlarging, and/or
  misaligning of the vertebrae
  (spondylolisthesis)
- degenerated discs
- compression of peripheral sections of
  the nerve, usually from soft tissue
  tension in the piriformis or
  related muscles
- significant misalignment of the sacral
  vertebrae over long time periods
- tumors impinging on the spinal cord
  or the nerve roots.
- trauma to the spine
- spinal stenosis caused by bone spurs,
  vertebral dislocation,
  inflammation, or herniated disc
- active trigger points of the lower back
  and the gluteus muscles occuring
  when muscles become ischemic
  due to injury or chronic muscular
  contraction

These physical initiating factors range from minor irritations of the nerves, to complete pinching and are accompanied with an inflammatory response, a sequence of aggravations, and can become debilitating.

The symptoms most often associated with sciatica are:

- Pain in lower back, buttock, and/or various parts of the leg and foot
- Numbness
- Sensation of cramping in legs
- Muscular weakness
- Pins and needles
- Tingling
- Difficulty moving or controlling the legs

The pain of sciatica runs from minor to severe and may be described as:

- Dull
- Aching
- Burning
- Shooting
- Sharp

Another important factor is that symptoms may be felt only on one side of the body. The pain of sciatica may be worsened by sitting, sneezing or coughing, straining at stool, bending or twisting.

Treatment of the underlying cause of the compression or irritation is often the most effective course. Most cases of sciatica are traditionally treated by one or more of the following:

- Chiropractic
- Anti-inflammatory medications (e.g. NSAIDs or oral steroids) though NSAIDs are frequently said to be no more effective than a placebo for acute sciatica.
- Non-surgical spinal decompression
- Physical therapy
- Surgery
- Stretching exercises
- Massage therapy

Homeopathy offers a combination of effective pain relief and support for the body's own healing mechanisms to work to their maximum capacity. Working with your health care team, you must decide as an individual, which approaches are best suited to your situation. Whether time alone, or medical intervention is necessary, homeopathy can be an important component of the healing program.

With an emergency acute situation at home, on the playing field, while hiking, etc. there are quick – one or two word summaries that help you choose a good homeopathic 'first aid' remedy. This type of summary in no way gives the full picture of the capability or uses of the remedy – but they are handy to know. A sample of these would be:

- Aconite for shock
- Arnica for trauma, bruising, hemorrhage
- Cantharis for burns
- Hypericum for any injury to nerve centers and tissues (including areas rich in nerves such as the fingers)
- Apis for acute synovitis
- Ledum for penetrating injuries

In homeopathic reference books, these quick summaries are known as 'keynotes'. While they are handy – they can also lead you to select too quickly, omitting consideration of a better remedy choice. Use 'keynotes' only in emergencies and try to expand your personal references whenever possible to find the best remedies for the acute situations you experience on a regular basis.

The following list of 19 remedies includes some of the keynotes most often associated with symptoms of sciatica.

Sciatic nerve

Arsenicum Album – Burning shooting, stitching, tearing pain, coldness in lower limbs, weakness, heaviness, twitching, spasms, uneasiness. Cramps in calves. especially for older people. Better from heat and warm drinks. Worse in wet weather, cold, cold drinks, and on the right side.

Belladonna – Severe neuralgic pains. Throbbing, shooting and stabbing pain. Lower back pain and pain in hips and thighs. Pain runs from head downward. Hot, burning, heavy and full sensations. Better with warm wraps, and firm pressure applied gradually. Worse for cold and drafts, lying down, touch, noise, with jarring motion and between 3pm and 9pm

Berberis vulgaris – Sharp stabbing pain, as if a thorn were stuck in the back. Stitching pain in the back and neck. Shooting pain down the leg. Deep pain in the lower back. Bubbling sensation in kidney area, over the skin, or in joints. Pain that moves from side to side. Better standing, Worse on the left side, for any motion, rising from sitting, twilight and at night.

Bryonia – Throbbing pain on motion, stiffness. Fullness and heaviness. Stitches and stiffness in small of back. Worse walking or turning, for slightest touch and from sudden changes in the weather. Better with stillness, firm pressure, coolness and open air.

Calcarea carbonica – Back feels too weak to sit upright in a chair. Vertebrae feel loose. Weakness and sinking sensation. Pain deep in muscles. Swollen joints. Worse in cold or wet weather, physical exertion, standing, lifting, stooping and worrying. Better in dry weather, lying still, lying on painful side, and any soothing with hands.

Causticum – Burning pains. Soreness, rawness, tearing pains in joints, bones, muscles and fibers. Bruised, darting pain in coccyx. Pain goes forward or to the thighs. Pain in hips, worse when coughing. Paralytic weakness. Better for cold drinks, damp , wet weather, warmth of bed. Worse for dry or cold air, extremes in temperature, stooping, exertion, 3 to 4 am, and riding in cars.

Colocynth – Severe pain. Cramp like pain in hip. Pain shoots down leg. Pain may be burning, digging, rending or tearing. Pain comes in waves. Cramping, twitching, constriction and contraction. Pain as if stabbed by a spear. Left sided. Drawing, tearing pain. Better for firm pressure and warmth, lying with knees drawn up to chest. Worse for gentle touch, resting on back, stooping, change in season, damp weather, cold winds.

Droscerea – Better when the spine is extended. Paralytic pains in the coxofemoral joint and thighs, all limbs feel lame. Sore, bruised pain. Worse for lying down and getting warm in bed.

Gelsemium – Dull, heavy, aching pain in muscles or bone. Pain runs up the back toward the head. General prostration. Slightest exertion is exhausting. Deep seated pain in lower back, generally passes upward, but may extend to hips and legs. Chilliness up and down the back. Better lying propped up by pillows and very still, pressure, fresh air, with stimulants. Worse when resting at night, before a storm, emotion or excitement, mid-morning (10 am), heat of the sun.

Hypericum – The classic remedy for ANY nerve damage or pain. Painfully sensitive spine. Pressure over sacrum. Violent pain and inability to walk after injury to coccyx. Shooting pain from injured spot. Tearing, shaking, rheumatic pains. Tingling, burning and numbness. Paralytic weakness. Prickling as from being stuck with needles. Better for bending head back or lying face down. Words for shock, exertion, jarring, light touch, change of weather, cold and damp.

Kali Carbonicum- Pains are sharp and cutting, throbbing or stitching. When sciatic pain is made worse by coughing, especially around 3 am. Small of the back feels weak. Burning in spine with sudden sharp pains extending down the back and to the thigh. Back feels as if broken. Stiffness and paralytic pain in back. Stitching pain in kidney region. Pain in buttocks, hip joints and thighs. Better for motion, warm humid weather, sitting with elbows on knees. Worse for cold air, drafts, early morning, and sudden motion.

Lycopodium – Burning pain, bubbling sensation, numbness, tearing pains, twitching and jerking. Better with motion, especially restlessness better with motion, warm drinks, cool air. Worse in the afternoon and when lying on the affected side, cold food or drink, sitting erect.

Magnesia Phosphorica – For the classical 'electric' shooting pains down the leg, worse on coughing and better for flexion of the spine. Feet very tender. Right sided. Neuralgic pain relieved by warmth.

Pulsatilla – Shooting pains, throbbing, pain in sacrum from sitting, wandering stitching pains. May be worse on right side. Sensations of pressure, congestion, constriction. Back feels bandaged. Better for consolation, breezes, cold applications, cold drink and cold food. Worse in the evening, with heat, wet weather, stooping.

Rhus toxicodendron – Pain and stiffness in the small of the back. Tearing pain in tendon, ligaments. Bones ache. Sciatic pain radiates from lower back and tears down the thighs. Better for motion or lying on something hard, stretching out limbs, warm weather, rubbing affected area. Worse for cold, wet, at night, while sitting, lying on back or right side.

Sepia – Weak, empty, hollow feeling in lower back. Sudden pain as if struck by a hammer. Bearing down pain in lower back which may hinder breathing. Better for warmth, vigorous motion, pressing back against something hard. Worse for cold air, dampness, and before thunderstorms.

Silica – Burning, weakness, sensitive to drafts on the back. Curvature of spine. Pain runs up the back toward the head. Spinal irritation. Better for warmth and humid weather. Worse for changes in weather, touch, pressure, riding in car, excitement.

Sulphur – Sensation as if vertebrai were gliding over one another. Stiffness. Curvature of spine. Lower back pain extends to stomach. Unsteady gait. Cannot walk erect, walks stop shouldered – can straighten up only after moving. Feebleness of tone. Better with warm applications and dry warm weather, lying on right side, hot drinks. Worse for warmth of bed, standing, stooping, jarring motion. Periodic (every 7 days)

Zinc – Burning along spine. Pain in small of the back. Can not bear to have back touched. Spinal irritation. Dull aching in spine, worse for sitting. Better for hard pressure, warm open air, motion, rubbing. Worse for physical or mental exertion, touch, noise.

REMEDY SUMMARIES

Out of over 3000 homeopathic remedies currently known and documented, there are literally hundreds that might be useful in the treatment of an individual with individuals with sciatic pain. We will focus on those best known for their affinity for acute situations. These few remedies are useful to alleviate suffering and speed the healing process – helping the individual pass the acute phase quickly and enter longer term healing and recovery. At that point, it is certainly recommended that a full homeopathic case may be taken and that homeopathic remedies may then be used in concert with appropriate therapies to insure the individual regains the best possible health.

A good materia medica will provide a full and complete description of the action of homeopathic remedies, but these can fill volumes. They describe the known actions and affinity of each remedy to mind and body carefully including symptoms and sensations. Depending on your needs and interests, you are encouraged to explore these fascinating books in bookstores or online by typing 'homeopathic material medica' in a search engine.

The following remedies will be presented in abbreviated form, including a very short overview, and a set of keynotes and modalities. These are drawn primarily from *Homeopathic Materia Medica* by William Boericke,

MD, originally published in 1901, with minor additions from a wide variety of additional sources.

Remedy Name	Common Abbreviation
Arsenicum album	Ars or Ars alb
Belladonna	Bell
Berberis vulgaris	Berb
Bryonia alba	Bry
Calcarea carbonica	Calc or Calc carb
Causticum	Caust
Colocynthis	Coloc
Droscerea rotundifolia	Dros
Gelsemium sempervirens	Gels
Hypericum perforatum	Hyper
Kalium carbonicum	Kali c. or Kali carb
Lycopodium clavatum	Lyc
Magnesium phosphoricum	Mag p. or Mag phos
Pulsatilla pratensis	Puls
Rhus toxicodendron	Rhus t or Rhus tox
Sepia officinalis	Sep
Silica terra	Sil
Sulphur	Sulph
Zincum metallicum	Zinc

Arsenicum Album

A profoundly acting remedy on every organ and tissue. Its clear-cut characteristic symptoms and correspondence to many severe types of disease make its homeopathic employment constant and certain. Its general symptoms often alone lead to its successful application. Among these the all-prevailing debility, exhaustion, and Restlessness, with nightly worse, are most important. Great exhaustion after the slightest exertion. This, with the peculiar irritability of fibre, gives the characteristic Irritable weakness. burning pains. Unquenchable thirst. Burning relieved by heat. Seaside complaints . Injurious effects of fruits, especially more watery ones. Gives quiet and ease to the last moments of life when given in high potency. Fear fright and worry. Green discharges). Arsenicum Album. should be thought of in ailments from alcoholism, Ptomaine poisoning, stings, dissecting wounds, chewing tobacco; ill effects from decayed food or animal matter; odor of discharges is putrid; in complaints that return annually. Anaemia and chlorosis. Degenerative changes. Gradual loss of weight from impaired nutrition. Maintains the system under the stress of malignancy regardless of location. Malarial cachexia. Septic infections and low vitality.

Selected Keynotes

Mind
- Feels vulnerable and defenseless esp. concerning disease and death.
- Needs company for reassurance and support.
- Demanding and clinging.
- Fastidious. Orderly
- Restlessness with anxiety and or weakness
- Thinks it useless to take medicine

Generalities

- Chilly
- Burning pains
- Right sided
- Periodicity (14 days)
- Weakness
- Alternating complaints

Back
- Stiffness of the back ascends from coccyx to nape.
- Cold creeps on back.
- As if warm air streaming up spine to the head.
- Bruised pain in small of back.
- Weak lumbar region.
- Sciatica ameliorated by walking and hot applications.
- Drawing in of shoulders
- Pain and burning in back

Extremities
- Paralytic weakness
- Black or blue nails.
- Coldness
- Twitching, trembling, spasms
- Weariness of limbs.
- Trembling of hands, of limbs.
- Tingling in fingers.
- Feet; weak, weary and numb, oedematous.
- Ulcers on soles, heals or toes.
- Wooden feeling in soles.
- Sore pain in ball of toes, while walking.
- Uneasiness in lower limbs, must move feet constantly or walk about.
- Toes bend downwards.
- Peripheral neuritis.
- Cramps in calves.
- Swelling of feet.
- Limbs heavy.
- Restless feet.
- Paraplegia, with atrophy.
- Paralysis, with contraction of limbs.

- Burning pains
- Peripheral neuritis

Modalities

BETTER from heat; from head elevated; warm drinks, motion, walking about, warm wraps, sitting erect, open air

WORSE, wet weather, after midnight; from cold, cold drinks, or food, exertion, Right side.

Belladonna

Belladonna acts upon every part of the nervous system, producing active congestion, furious excitement, perverted special senses, twitching, convulsions and pain. It has a marked action on the vascular system, skin and glands. Belladonna always is associated with hot, red skin, flushed face, glaring eyes, throbbing carotids, excited mental state, hyperaesthesia of all senses, delirium, restless sleep, convulsive movements, dryness of mouth and throat with aversion to water, Neuralgic pains that come and go suddenly. Heat, redness, throbbing and burning. great children's remedy. Epileptic spasms followed by nausea and vomiting. Scarlet fever and also prophylactic. Here use the thirtieth potency. Corresponds to the symptoms of "air-sickness" in aviators. Give as preventive. No thirst, anxiety of fear. Belladonna stands for violence of attack and suddenness of onset.

Selected Keynotes

Great intensity of symptoms.

Mind
- Vivid emotions, imaginations
- Senses acute. Sensitive to light, noise, jarring
- Irritability. Sudden fits of anger.

Generalities
- Robust, vital, plethoric.
- Pulsating pains - Pains (esp. neuralgic) come and go suddenly.
- Convulsions, twitches.
- Mucous membranes become dry.
- effects are sudden and violent.
- Burning heat, bright redness and dryness
- Severe Neuralgic Pain, that comes and goes suddenly.

- Pain is throbbing, sharp, cutting or shooting, and is of maddening severity; coming and going in repeated attacks.
- Acuteness of all senses
- Changeableness.
- Spasms described as shocks, jerks and twitching

Back
- Stiff neck and shoulder (right).
- Back feels broken.
- Lumbago, with pain in hips and thighs.
- Pressure on dorsal region most painful
- Cramp - intermittent pulling (head bent back)
- Pain with movement or jarring
- Stiffness when rising from a seat

Extremities
- Jerks, spasms or trembling in limbs.
- Joints, swollen, red, shining; with red streaks radiating.
- Heaviness and paralytic feeling in limbs.
- Cold extremities.
- Involuntary limping.
- Weakness and tottering gait.
- Shooting pains along limbs
- Shifting rheumatic pains

Modalities

BETTER sitting semi-erect or standing erect, warm room

WORSE for touch, jarring motion, noise, draft, after noon, lying down, heat, sun, sudden change in temperature

Berberis vulgaris

Rapid change of symptoms-pains change in regard to place and character-thirst alternates with thirstlessness, hunger, and loss of appetite, etc. Acts forcibly on the venous system, producing pelvic engorgements and haemorrhoids. Hepatic, and rheumatic affections, particularly with urinary, haemorrhoidal and menstrual complaints. Old gouty constitutions. Pain in region of kidneys is most marked; hence its use in renal and vesical troubles, gall-stones, and vesical catarrh. It causes inflammation of kidneys with haematuria. Pains may be felt all over body, emanating from small of back. It has also marked action on the liver, promoting the flow of bile. Often called for in arthritic affections with urinary disturbances. Wandering, radiating pains. Acts well in fleshy persons, good livers, but with little endurance. Spinal irritation. All Berberis pains radiate, are not worse by pressure, but worse in various attitudes, especially standing and active exercise.

Common name: Barberry, Berberidaceae

Selected Keynotes

Mind
- Apathetic
- Indifferent

Generalities
- mentally and physically tired- not inclined to do anything.
- Prematurely old and worn out men and women.
- Pains may be sharp, twinging, and radiating in an uncoordinated fashion (no pattern)
- Pains rapidly change their locality and character.
- Pains radiate from one point; shooting outward

or all over; sticking; burning; sorness.
- Numbness; insensible to extreme heat and cold.
- Cold feeling in the bones

Back
- Stitches in neck and back
- Backache, with severe prostration
- Spinal irritation
- Sore spot under scapula (usually right side).
- Crushing, stitching, paralyzing pain, in lumbar region.
- Post-operative pain in lumbar region.

Extremities
- Outer side of the thighs cold, ulcerative pain in heels.
- Pain in balls of feet on stepping.
- Intense weariness and lameness of legs after walking a short distance.
- Stitches between metatarsal bones as from a nail
- Wandering radiating pains
- Sticking pain in region of kidneys radiating thence around abdomen to hips and groins
- Numb, bruised sensation
- Sensation of cold on outside of thighs

Modalities

WORSE, motion, standing after sitting, lying down, sitting, jarring movement

Bryonia

Acts on all serous membranes and the viscera they contain. Aching in every muscle. The general character of the pain here produced is a stitching, tearing; worse by motion, better rest. These characteristic stitching pains, greatly aggravated by any motion, are found everywhere, but especially in the chest; worse pressure. Mucous membranes are all dry. The Bryonia Alba patient is irritable; has vertigo from raising the head, pressive headache; dry, parched lips, mouth; excessive thirst, bitter taste, sensitive epigastrium, and feeling of a stone in the stomach; stools large, dry, hard; dry cough; rheumatic pains and swellings; dropsical effusions into synovial and serous membranes. Bryonia Alba affects especially the constitution of a robust, firm fiber and dark complexion, with tendency to leanness and irritability. It prefers the right side, the evening, and open air, warm weather after cold days, to manifest its action most markedly. Physical weakness, all-pervading apathy. Complaints apt to develop slowly.

Common names: White Bryony, Wild Hop.
Cucurbitaceae

Selected Keynotes

Mind
- Aversion to being disturbed.
- Irritable, wants to be left alone. Reserved.
- Thoughts, talks about business. Determined.
- Despair of recovery. Seem to hold others responsible for their suffering.
- Dullness morning on waking,

Generalities
- DRYNESS mucous membranes
- Often slow onset of acute diseases.
- Injuries, sprains, etc.

- Stitching pains.
- Slow onset
- Great thirst for large quantities
- Also possible: dry mouth and thirst less

Back
- Painful stiffness in nape of neck
- Stitches and stiffness in small of back

Extremities
- Rheumatism of joints and muscles, arthritis, gout
- Inflammation of joints with swelling and stiffness.
- Pain that feels like bone in hurting

Modalities

BETTER, lying on Painful side, pressure, rest, cold applications.

WORSE, warmth, any motion, morning, eating, heat, hot weather, jarring, exertion, touch. Cannot sit up- gets faint and sick.

Calcaria carbonica

*This great Hahnemannian anti-psoric is a constitutional remedy
Par Excellence. Its chief action is centered in the vegetative
sphere, impaired nutrition being the keynote of its action, the
glands, skin and bone, being instrumental in the changes
wrought. Increased local and general perspiration, swelling of
glands, scrofulous and rachitic conditions generally offer
numerous opportunities for the exhibition of Calcarea
Carbonica. Incipient phthisis. It covers the tickling cough,
fleeting chest pains, nausea, acidity and dislike of fat. Gets out
of breath easily. A jaded state, mental or physical, due to
overwork. abscesses in deep muscles; polypi and exostoses.
Pituitary and thyroid disfunction. Is a definite stimulant to the
periosteum. Easy relapses, interrupted convalescence. Persons of
scrofulous type, who take cold easily, with increased mucous
secretions, children who grow fat, are large-bellied, with large
head, pale skin, chalky look, the so-called leucophlegmatic
temperament; affections caused by working in water. Great
sensitiveness to cold; partial sweats.*

Source: Oyster shell

Common name: Calcium carbonate

Selected Keynotes

Mind
- Hard working, overworking, capable, conscientious,
over-responsible, take on too much.
- Practical, down to earth. Needs security.
- Obstinate.
- A fear of the unknown. Many small fears. Worry
  about small things.
- Anxiety about health, future. Despair of
  recovery.
- Sadness. Melancholy.

Generalities
- Pillow wet from perspiration neck.
- Chilly
- Sour odor of discharges, of body.
- Bones: Weak; disturbed development, caries.

Back
- Weakness of back.
- Curvature of spine.
- Lumbago, sciatica.
- Swelling of cervical glands.

Extremities
- Weakness. Sprains easily.
- Arthritis, rheumatism
- Cold, clammy hands and feet.
- Nails brittle
- Cramps in calves at night in bed.

Modalities

BETTER, dry climate and weather; lying on painful side. Heat of the sun.

WORSE, from exertion, mental or physical; ascending; cold in every form; water, washing, moist air, wet weather; during full moon; standing.

Causticum

Manifests its action mainly in chronic rheumatic, arthritic and paralytic affections, indicated by the tearing, drawing pains in the muscular and fibrous tissues, with deformities about the joints; progressive loss of muscular strength, tendinous contractures. Broken down seniles. In catarrhal affections of the air passages, and seems to choose preferable dark-complexioned and rigid-fibered persons. Restlessness at night, with tearing pains in joints and bones, and faint-like sinking of strength. This weakness progresses until we have gradually appearing paralysis. Local paralysis, vocal cords, muscles of deglutition, of tongue, eyelids, face, bladder and extremities. The skin of a Caucasian person is of a dirty white sallow, with warts, especially on the face. Emaciation due to disease, worry, etc., and of long standing. Burning, rawness, and soreness are characteristic.

Source: Causticum Hahnemanni – Potassium Hydrate

Selected Keynotes

Mind
- Sensitive, excitable, perceptive individuals.
- Sympathetic. Full of cares. Anxiety about others.
- Strong sense of justice. Idealism.
- History of many shocks, griefs, disappointments.
- Ailments from fright, prolonged grief.
- Mind gets tired, weak, breaks down.

Generalities
- Complaints of nervous and musculo-skeletal system.
- Chilly.

Nervous system
- Tics, chorea, twitching, convulsions, restlessness.
- Degenerative neurological diseases: e.g. MS.
- Paralysis esp. right side.
- Numbness.

Back
- Stiffness.
- Torticollis.
- Sciatica

Extremities
- Gradual paralysis
- Ataxia.
- Contractions, indurations of tendons
- Carpal tunnel syndrome
- Arthritis of small joints
- Restless legs at night, during sleep.
- Twitchings
- Paralysis, pain right deltoid.
- Pain in hamstrings, with limited extension
- Contractures of tendons, stiff neck
- Stiffness around knees and ankles with
  tightness of the hamstrings

Modalities

BETTER, in damp, wet weather; warmth. Heat of bed

WORSE, dry, cold winds even in Clear fine weather, Cold air; from motion of carriage.

Colocynthis

Develops most of its symptoms in the abdomen and head, causing intense neuralgias. It is especially suitable for irritable persons easily angered, and ill effects therefrom. Persons with a tendency to corpulency. The neuralgic pains are nearly always relieved by pressure. Cramps and twitching and shortening of muscles. Constrictions and contractions. Urinous odor or perspiration . Agonizing pain in abdomen, causing patient to bend double, is most characteristic. Sensations: Cutting, twisting, grinding, contracting and bruised; As if clamped with Iron Bands.

Selected Keynotes

Mind
- Anger, irritable, quarrelsome, with indignation
- Symptoms follow intense emotions
- Restlessness
- Nervousness
- Becomes angry when questioned

Back
- Lumbar pain is piercing, stabbing, pulling
- On movement, pain is sudden, stitching, sharp, or shooting
- Piercing or stabbing pain while lying down
- Pain extends down right thigh
- Muscles and tendons feel too short
- Numbness with pains

Extremities
- Pain in left knee joint.
- Cramps in leg and calf
- Pain in hip
- Excessive foot perspiration

- Weakness in knee while walking
- Contraction of the muscles
- Stiffness of joints and shortening of tendons
- Spontaneous luxation of the hip-joints
- Contraction of muscles

Modalities

BETTER, doubling up, hard pressure, warmth, lying with head bent forward.

WORSE, gentle touch, from anger and indignation

Drosera rotundifolia

Affects markedly the respiratory organs and was pointed out by Hahnemann as the principal remedy for whooping-cough. Drosera can break down resistance to tubercle and should therefore be capable of raising it. Laryngeal phthisis is benefited by it. Phthisis pulmonum; vomiting of food from coughing with gastric irritation and profuse expectoration. Pains about hip-joint. Tubercular glands. Spasmodic cough.

Common name: Sundew

Selected Keynotes

Mind
- Angry at trifles.
- Difficult concentration.
- Anxiety, sadness when alone.
- Restlessness

Generalities
- Bed feels too hard
- Chill in bed

Back
- Sticking, lancinating pains

Extremities
- Limbs feel cold
- Lameness
- Paralytic pains in the coxofemoral joint and Thighs
- Laming pain in right hip joint and thigh with pain in ankle; must limp when walking.
- Limbs feels sore

- Pain in the long bones.
- Pain in humerus at night only.

Modalities

WORSE, heat, emotional excitement, after midnight, lying down, on getting warm in bed, drinking, singing, laughing.

BETTER, open air, pressure

Gelsemium

sempervirens

Centers its action upon the nervous system, causing various degree of Motor paralysis. General prostration. Dizziness, drowsiness, dullness, and trembling. Slow pulse, tired feeling, mental apathy. Paralysis of various groups of muscles about the eyes, throat, chest, larynx, sphincter, extremities, etc. Post-diphtheritic paralysis. Muscular weakness. Complete relaxation and prostration. Lack of muscular coordination. General depression from heat of sun. Sensitive to a falling barometer; cold and dampness brings on many complaints. Sluggish circulation. Nervous affections of cigar makers. Influenza. Measles. Pellagra.

Common name: Yellow jasmine

Selected Keynotes

Mind
- Feeling of weakness, of not being able to cope with daily life, responsibilities, work.
- Timidity.
- Desire to be quiet, to be left alone. Avoid people and distress of life.
- Fear of falling, crowds.
- Mental weakness. Dullness. Forgetful.

Generalities
- Weakness, paresis, paralysis
- Ailments from anticipation, fright, excitement, bad news
- Trembling from weakness, fear, anticipation .
- Flushes of heat alternating with chills, -

Back
- Chilliness up and down the back
- Weakness.
- Pain and stiffness cervical region extending to head.
- Pain under left scapula.

Extremities
- Loss of power of muscular control.
- Paralysis. Contractions.
- Trembling and weakness
- Heavy feeling, especially lower limbs.

Modalities

BETTER, bending forward, by profuse urination, open air, continued motion, stimulants.

WORSE, damp weather, fog, before a thunderstorm, emotion, or excitement, Bad news, tobacco-smoking, when thinking of his ailments; at 10 A.m.

Hypericum

perforatum

The great remedy for injuries to nerves, especially of fingers, toes and nails. Crushed fingers, especially tips. Excessive painfulness is a guiding symptom to its use. Prevents lockjaw. Punctured wounds. Relieves pain after operations. Spasms after every injury. Has an important action on the rectum; hemorrhoids. Coccydynia. Spasmodic asthmatic attacks with changes of weather or before storms, better by copious expectoration. Injured nerves from bites of animals. Tetanus. Neuritis, tingling, burning and numbness. Constant drowsiness.

Common name: St. John's Wort

Selected Keynotes

Injuries to nerves and spine.

Mind
- Symptoms after injuries include: depression, dullness, forgetfulness, hysteria
- Feels as if lifted in the air.
- Fear of falling from height, downward motion

Generalities
- INJURIES TO PARTS RICH IN NERVES
- Pain: shooting upward along nerve
- Convulsions after injury, esp. to spine, head
- Ailments of shock, fright
- Phantom pains

Back
- Injuries to spine and coccyx
- Pain shoots up the spine and down limbs.
- Bruised coccyx after labor.

Extremities
- Neuritis.
- Injuries to finger tips.
- Rheumatic pains.

Modalities

BETTER, bending head backward.

WORSE, in cold; dampness, in foggy weather; before thunderstorm; in close room; least exposure; touch.

Kalium carbonicum

The weakness characteristic of all Potassium Salts is seen especially in this, with soft pulse, coldness, general depression, and very characteristic stitches, which may be felt in any part of the body, or in connection with any affection. All Kali pains are sharp and cutting; nearly all better by motion. Never use any Salts of Potash where there is fever (T. F. Allen). Sensitive to every atmospheric change, and intolerance of cold weather. Early morning worse is very characteristic. Fleshy aged people, with dropsical and paretic tendencies. Sweat, backache, and weakness. Throbbing pains. Tendency to dropsy. Tubercular diathesis. Pains from within out, and of stinging character. "Giving-out" sensation. Fatty degeneration. Stinging pains in muscles and internal parts. Twitching of muscles. Pain in small spot on left side. Hypothyroidism. Coxitis.

Common Name: Potassium Carbonate

Selected Keynotes:

Mind
- Rigid, inflexible, dogmatic
- Systematic, proper, routine-oriented.
- Can be sensitive but will not show it.
- Emotions, anxiety felt in solar plexus
- Hypersensitive to touch, ticklish.
- Irritability at trifles.

Generalities
- Chilly
- Stitching pains

Back
- Weakness. Lumbago, sciatica.
- Stiffness and paralytic feeling
- Pain in thighs and hip joint
- Sharp pains extending up and down back and to thighs
- Pain from hip to knee

Extremities
- Back and legs give out
- Heaviness and tearing in limbs
- Deformative arthritis,
- Dropsies.
- Sensitive footsoles

Modalities

BETTER in warm weather, during the day

WORSE for cold, wet, draft, from soup and coffee, lying on painful side

Lycopodium clavatum

In nearly all cases where Lycopodium is the remedy, some evidence of urinary or digestive disturbance will be found. Lycopodium is adapted more especially to ailments gradually developing, functional power weakening, with failures of the digestive powers, where the function of the liver is seriously disturbed. Atony. Malnutrition. Mild temperaments of lymphatic constitution, with catarrhal tendencies; older persons, where the skin shows yellowish spots, earthy complexion, uric acid diathesis, etc; also precocious, weakly children. Symptoms characteristically run from right to left, acts especially on right *side of body, and are worse from about 4 to 8 pm. In kidney affections,* red *sand in urine, backache, in renal region; worse before urination. Intolerant of cold drinks;* craves everything warm. *Best adapted to persons intellectually keen, but of weak, muscular power. Deep-seated, progressive, chronic diseases. Carcinoma. Emaciation. Debility in morning. Marked regulating influence upon the glandular (sebaceous) secretions. Pre-senility. Ascites, in liver disease. Lycop patient is thin, withered, full of gas and dry. Lacks vital heat; has poor circulation, cold extremities. Pains come and go suddenly. Sensitive to noise and odors.*

Common Name: Club moss

Selected keynotes

Mind
- Often intellectual people like schoolteachers, lawyers.
- Timid
- Presents themselves different than they feel Inside
- Extremely sensitive

Back
- Burning between scapulae as of hot coals.
- Pain in small of back
- Torticollis
- Lumbago
- Sciatica ameliorated by walking
- Sciatica, worse right side.

Extremities
- Numbness, also drawing and tearing in limbs, especially while at rest or at night.
- Heaviness of arms.
- Tearing in shoulder and elbow joints.
- One foot hot, the other cold.
- Chronic gout, with chalky deposits in joints.
- Profuse sweat of the feet.
- Pain in heel on treading as from a pebble.
- Painful callosities on soles;
- Toes and fingers contracted.
- Cannot lie on painful side.
- Hands and feet numb.
- Right foot hot, left cold.
- Cramps in calves and toes at night in bed.
- Limbs go to sleep.
- Twitching and jerking.

Modalities

BETTER, by motion, after midnight, from warm food and drink, on getting cold, from being uncovered.

WORSE, right side, from right to left, from above downward, 4 to 8 pm; from heat or warm room, hot air, bed. Warm applications, except throat and stomach which are better from warm drinks.

Magnesium phosphoricum

The great anti-spasmodic remedy. Cramping of muscles with radiating pains. Neuralgic pains relieved by warmth. Especially suited to tired, languid, exhausted subjects. Indisposition for mental excertion. Guitre.

Selected Keynotes

Mind
- Oversensitive.
- Lamenting all the time about pain.
- Irritable.
- Capricious.
- Indisposition to mental exertion.

Generalities
- Neuralgic pains
- General muscular weakness

Neck and back
- Stiffness of neck and back
- Cramps
- One vertebra seems absent

Extremities
- Cramps from prolonged exertion.
- Sciatica,
- Weakness upper limbs, trembling. Parkinson.
- Cramping pains
- Cramps in calves

- Spasmodic pain
- Acute stabbing pain
- Boring pain

Modalities

BETTER for warmth, bending double, pressure, friction.

WORSE on right side, COLD, touch, night.

Pulsatilla

The weather-cock among remedies. The disposition and mental state are the chief guiding symptoms to the selection of Pulsatilla. It is pre-eminently a female remedy, especially for mild, gentle, yielding disposition. Sad, crying readily; weeps when talking; Changeable, contradictory. The patient seeks the open air; always feels better there, even though he is chilly. Mucous membranes are all affected. Discharges thick, bland, and yellowish-green. Often indicated after abuse of iron tonics, and after badly-managed measles. Symptoms ever changing. thirstless, peevish, and chilly. Great sensitiveness. Wants the head high. Feels uncomfortable with only one pillow. Lies with hands above head.

Common Name: Anemone. Ranunculaceae

Selected Keynotes

Mind
- Forsaken feeling, often expressed in the need for attention. They do anything to fulfill this need.
- Often mild, but can be irritable.
- Changeable moods.
- Timidity, blushes easily.

Generalities
- Changeable symptoms
- Warm-blooded
- Wandering pains

Back
- Sensation of cold water poured down the back
- Pain in back and small of back as from prolonged stooping.
- Curvature of the upper part of the spine.
- Shooting pain in the nape and back.
- Back feels bandaged.
- Pain in sacrum on sitting.

Extremities
- Wandering arthritis
- Hot feet – uncovers them at night
- Pains in the heels
- Rapidly shifting pains in limbs
- Joints swollen and red
- Dorsum of foot swollen, oedematous
- Pains down lower limbs, alternating sides
- Legs heavy
- Cold sweat on legs
- Foul foot sweat
- Veins full, varicose, painful
- White swelling of knee
- Inclination to stretch the feet
- Numbness in legs when remaining long standing

Modalities

BETTER In the open air; lying on painful side; cold air or cool room; eating or drinking cold things; cold applications, consolation

WORSE In a warm close room; evening, at twilight; on beginning to move; sun; lying on the left, or on the painless side; very rich, fat, indigestible food; pressure on the well side if it be made toward the diseased side; warm applications; heat

Rhus toxicondendron

The effects on the skin, rheumatic pains, mucous membrane affections, and a typhoid type of fever, make this remedy frequently indicated. Rhus Toxicodendron affects fibrous tissue markedly - joints, tendons, sheaths - aponeurosis, etc., producing pains and stiffness. Post-operative complications. Tearing asunder pains. Motion always "limbers up" the Rhus patient, and hence he feels better for a time from a change of position. Ailments from strains, overlifting, getting wet while perspiring. Septic conditions. Cellulitis and infections, carbuncles in early stages . Rheumatism in the cold season. Septicaemia.

Common name: Poison ivy

Selected Keynotes

Mind
- Busy, restless
- Timid and mild, yet lively.
- Later on irritable, impatient.
- Depressed, morose.
- Becomes rigid and stiff as with the muscles and joints.
- Behavior disorders with restlessness, irritability.

Generalities
- Sudden periodic chills esp. at night.
- Parkinson's disease.

Back
- Pain and stiffness
- Stiffness of the neck,

Extremities
- Arthritis, rheumatism with painful stiffness
- Pain Left shoulder , inner side left scapula.
- Restless extremities. Restless legs in bed at

night.
- Sprains. Overexertion.
- Bursitis. Tendonitis.
- Chorea. Twitching. Tics.
- Joint pains and fevers

Overstretching or overexertion leaves muscles inflamed, stiff and in need of gentle motion and warmth

Sore muscles, bruised feeling, strained back muscles. Needs to stretch and move. Massage neck constantly. Muscles bound up in neck.

Muscles are painful and stiff

Modalities

BETTER, warm, dry weather, motion, walking, change of position, rubbing/massage, warm applications, from stretching out limbs.

WORSE, overexertion, during sleep, cold, wet rainy weather and after rain, foggy weather, thunderstorms, at night, during rest, draft, when lying on back or right side.

Sepia

*Acts specially on the portal system, with venous congestion.
Stasis and thereby ptosis of viscera and weariness and misery.
Weakness, yellow complexion, bearing-down sensation, especially
in women, upon whose organism it has most pronounced effect.
Pains extend down to back, chills easily. Upward tendency of its
symptoms. Easy fainting. "Ball" sensation in inner parts. Feels
cold even in warm room. Pulsating headache in cerebellum.*

Source: Ink of Cuttlefish

Selected Keynotes

Mind
- Irritable and easily offended
- Indolent
- Weeps when telling symptoms

Generalities
- periodicity – 28 days

Back
- Weakness in small of back
- Pains extend into back
- Coldness between shoulders
- Cramps in buttocks when stretching out legs
- Sudden back pain, as if struck by a hammer

Extremities
- Lower extremities lame and stiff
- Heaviness and bruised feeling
- Restlessness in all limbs
- Twitching and jerkings night and day

- Pain in heel
- Coldness in legs and feet

Modalities

BETTER, by exercise, hard pressure, warmth of bed, hot applications, motion, drawing limbs up, cold bathing, after sleep.

WORSE, forenoons and evenings; washing, jarring, stooping, dampness, left side, after sweat, cold air, before thunder-storm.

Silica

Imperfect assimilation and consequent defective nutrition. It goes further and produces neurasthenic states in consequence, and increased susceptibility to nervous stimuli and exaggerated reflexes. Diseases of bones, caries and necrosis. Organic changes; it is deep and slow in action. Periodical states; abscesses, quinsy, headaches, spasms, epilepsy, feeling of coldness before an attack. Ill effects of vaccination. Suppurative processes. It is related to all pustulous burrowings. Ripens abscesses since it promotes suppuration. Silica patient is cold, chilly, hugs the fire, wants plenty warm clothing, hates drafts, hands and feet cold, worse in winter. Lack of vital heat. Prostration of mind and body. Great sensitiveness to taking cold. Intolerance of alcoholic stimulants.

Source: flint

Selected Keynotes

Mind
- Want of self-confidence.
- Complaints from long mental exertion.

Generalities
- Chilly
- Stitching pains, splinter-like
- Frequent colds and infections with enlarged
  glands, that may suppurate.
- Weakness, lack of energy.
- Sensitivity
- Diseases of bone.

Back
- Diseases of bone and spine - Scoliosis.
  Curvatures. Caries.
- Weakness.

- Lumbago.
- Sciatica.
- Coldness. Sensitive to draft.
- Ailments from injury to the spine
- Irritation of the spine

Extremities
- Coldness.
- Nails break easily
- Pain in ankles and wrists
- Bunions
- Profuse, offensive, corrosive foot sweat.
- Arthritis

Pain arising from neglected injuries

Modalities

BETTER for warmth, wrapping up head, summer; warm food and drinks, in wet or humid weather.

WORSE in morning, from washing, uncovering, lying down, damp, lying on left side, cold, weather changes, draft, mental or physical exertion. In acute situations, heat may aggravate.

Sulphur

This is great Hahnemannian anti-psoric. Its action is centrifugal-from within outward-having an elective affinity for the skin, where it produces heat and burning, with itching; made worse by heat of bed. Inertia and relaxation of fiber; hence feebleness of tone characterizes its symptoms. Ebullitions of heat, dislike of water, dry and hard hair and skin, red orifices, sinking feeling at stomach about 11 am, and cat-nap sleep; always indicate Sulphur homeopathically. Standing is the worst position for Sulphur patients, it is always uncomfortable. Aversion to being washed. When carefully-selected remedies fail to act, especially in acute diseases, it frequently arouses the reactionary powers of the organism. Complaints that relapse. General offensive character of discharge and exhalations. Very red lips and face, flushing easily. Often great use in beginning the treatment of chronic cases and in finishing acute ones.

Common Names: Brimstone; Flowers of Sulphur.
The Element

Selected Keynotes

Mind
- Self-conscious.
- Critical
- Philosophical/intellectual type
- Practical type
- Aversion to dirt from others
- Fastidiousness may be seen also
- Obstinate and irritable.
- Hypochondriacal.

Generalities
- Warm-blooded but sometimes chilly
- Discharges offensive
- Profuse and offensive perspiration.
- Burning pains
- Periodicity, 7 days

Back
- Weakness and pain,
- Can't sit straight up or stand
  erect
- Drawing pain between shoulders
- Stiffness of nape
- Sensation as if vertebrae glided over
  each other

Extremities
- Feet hot and burning. Uncovers them at
  night.
- Offensive perspiration of feet.
- Left shoulder pain
- Heaviness – paretic feeling
- Rheumatic gout with itching
- Arthritis
- Brittle and distorted nails.

Modalities

BETTER Dry, warm weather, lying on the right
side, cold application,

WORSE At rest, heat, when standing, warmth
in bed, washing, bathing, changeable weather,
winter

Zincum metallicum

The provings picture cerebral depression. Tissues are worn out faster than they are repaired. Poisoning from suppressed eruptions or discharges. The nervous symptoms of most importance. Defective vitality. Impending brain paralysis. Period of depression in disease. Spinal affections. Twitchings. Pain, as if between skin and flesh. Great relief from discharges. Chorea, from fright or suppressed eruption. Convulsions, with pale face and no heat. Marked anaemia with profound prostration. It causes a decrease in the number, and destruction of red blood corpuscles. In chronic diseases with brain and spinal symptoms, trembling, convulsive twitching and fidgety feet are guiding symptoms.

Common Name: Zinc

Selected Keynotes

Mind
- Restless, hyper-active
- Oversensitive
- Over-active mind and ideas abundant
- Loquacity
- Irritability from noise, talking of others
- Fits of anger with weakness afterwards
- Brain and nerves are fatigued

Back
- Weakness - tired feeling in spine
- Pain, coccyx
- Spinal irritation
- Burning pain in spots
- Burning along whole spine

- Aching in spine; small of the back
- Spine sensitive, can not bear it touched.
- Cutting pain in inter-scapular region
- Dull aching pain about last dorsal or first
  lumbar vertebrae.
- Weakness in lumbar region

Extremities
- Restlessness, constantly moves legs and feet
- Numbness and formication in legs when lying.
- Twitches, jerks, electric-like activity in the
  muscles.
- Hyperactivity in the muscles.
- Painless paralysis.
- Pains in the limbs on becoming heated.
- Transverse pain esp. in upper limbs.
- Cramps, when cold.
- Toes feel swollen.
- Soles sensitive.
- Bones of the feet feel broken.
- Stumbling, spastic gait; totters while walking
- Painful varicoses; chronic.
- Burning in tibia.
- Paralysis of feet.
- Ulcerative pain in heel

Modalities

BETTER, rubbing and pressure, motion, warm open air

WORSE, cold, from touch, sitting, turning in bed,
stooping, standing, exhaustion (mental or physical)

Notes

Notes

Notes

Notes

www.ingramcontent.com/pod-product-compliance
Lightning Source LLC
Chambersburg PA
CBHW031251280526
45784CB00004B/1806